Georg Borowski

Merkwürdige und nützliche Thiere in Abbildungen

Borowski, Georg

Merkwürdige und nützliche Thiere in Abbildungen

Reihe: Historical Science, Band 41

ISBN: 978-3-86741-513-2

Auflage: 1
Erscheinungsjahr: 2010
Erscheinungsort: Bremen, Deutschland

© Europäischer Hochschulverlag GmbH & Co KG, Fahrenheitstr. 1, 28359 Bremen (www.eh-verlag.de). Alle Rechte beim Verlag und bei den jeweiligen Lizenzgebern.

Bei diesem Titel handelt es sich um den Nachdruck eines historischen, lange vergriffenen Buches. Da elektronische Druckvorlagen für diese Titel nicht existieren, musste auf alte Vorlagen zurückgegriffen werden. Hieraus zwangsläufig resultierende Qualitätsverluste bitten wir zu entschuldigen.

Georg Borowski

Merkwürdige und nützliche Thiere in Abbildungen

Simia Satyrus Linn.
Le Jocko.
Der Orang outang der kleinere.
The man of the wood.

Simia Faunus Linn.
Malbrouc.
Der Löwenschwanz.

Simia Mormon Alstroem.
Der Choras.
Tufted ape.

Lemur Catta Linn.
Mococo.
Eichhorn Affe.
Ring-tail Maucauco.

Vespertilio Vampyrus Linn.
La Roussette.
Fliegender Hund.
The Ternate Bat.

Bradypus Tridactylus Linn.
Paresseux.
Dreifingriges Faulthier.
Sloth.

D. Sotrmann del.
ex Museo Blochiano

Glasbach sc.

Myrmecophaga jubata Linn.
Tamanoir.
Grosser Ameisenfresser.
Great ant-eater.

D. Sotzmann del. — Glasbach sc.

Manis pentadactyla Linn.
Pangolin.
Fünffingriges Schuppenthier.
Short-tailed manis.

Dasypus sexcinctus Linn.
Encoubert. Buff:
Sechsgürtliche Armadill.
Six Banded armadillo.

Rhinoceros unicornis Linn.
Rhinoceros
Nashorn.
Rhinoceros

Elephas maximus Linn.
Elephant.

Trichecus Rosmarus Linn.
Morse.
Wallross.
Artic Wallrus.

Phoca ursina Linn.
Ours marin.
Der See=Bär.
The ursine Seal.

Canis aureus Linn.
Chacal.
Der Goldwolf.
The Jackal.

Canis Hyæna Linn.
L'Hyaene.
Das Grabthier.
Striped Hyaena.

Felis Leo Linn.

Lion.

Der Löwe.

The Lion.

D. Schreber del.   Glaßb. sc.

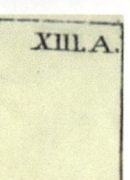

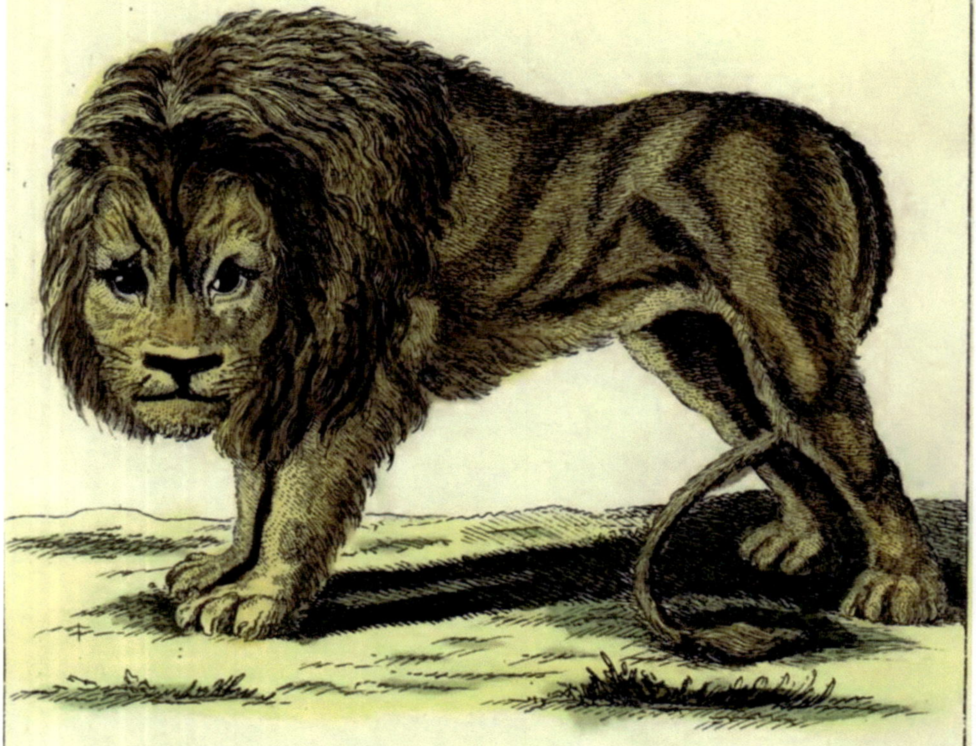

Felis Tigris Linn.
Le Tigre.
Das Tigerthier.
The Tiger.

Mustela Zibellina Linn:
La Zibeline.
Der Zobel.
The Sable.

Ursus Gulo Linn:
Le Glouton
Vielfraß.
The Glutton.

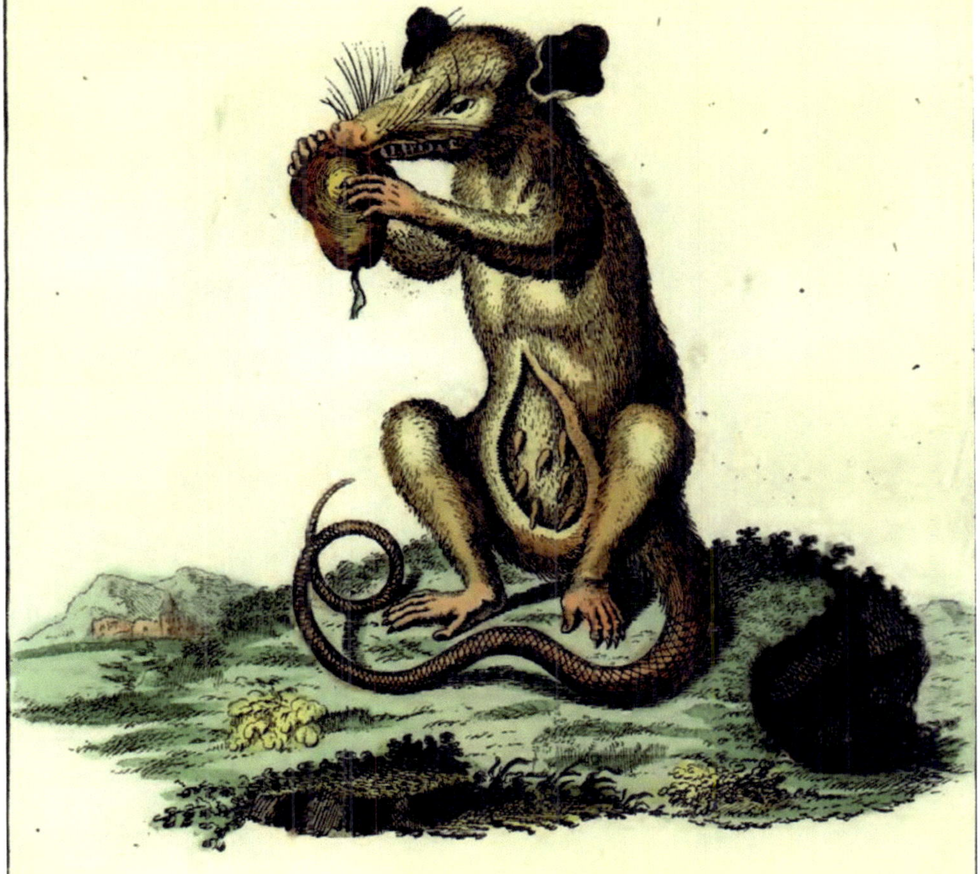

Didelphys Marſupialis Linn:

Le Sarigue

Beutelratze

The Virginian Oposſum

D. Solam: del:

Glab: fc

Talpa Asiatica Linn:
Taupe dorée
Goldmaulwurf
The Sibirian Mole

Sorex moschatus Pall.
Le Desman
Die Biesamratte
The Longnosed Beaver.

Erinaceus Auritus Pall.
Herisson de Siberie.
Langöhrige Igel.
The Sibirien Hedge-hog)

XXI.

Hystrix dorsata Linn:
L'Urson
Das verlarffte Stachelschwein.
The Canada Porcupine

XXII.

Lepus Angorensis Lin:
Le Lapin d'Angora.
Das Angorische Kaninichen.
The Angora Rabbet.

Castor Zibethicus Linn:
L'Ondatra.
Die Bisamratte.
The Musk Beaver.

Cavia Aguti Erxl.
L'Agouti.
Das Ferkelkaninchen
The Longnosed Cavy.

Mus amphibius Linn.
Le Rat d'eau blanc.
Die Wasserratze.
The Water Rat.

Glis Lemmus Lin:
Le Leming.
Die Norwegische Bergmaus. Lemming.
The Lappland Marmot

Iaculus Orientalis Erxl.
Le Gerbo.
Die Ægyptische Bergratte.
The Ægyptian Jerboa.

Sciurus Striatus Linn:
L'Ecureuil Suisse.
Das gestreifte Eichhorn.
The ground Squirrel

Camelus Glama Linn:
Le Chameau du Perou
Das Peruanische Schafkameel.
The Glama

Moschus Americanus Erxl:
Le Chevrotain de Surinam.
Das Surinamsche Hirschgen.
The Brasilian Musk.

Cervus Camelopardalis Lin:
La Giraffe.
Die Giraffe. Kameelparder.
The Camelopard.

Cervus Tarandus Linn:
Le Renne.
Das Rennthier.
The Rein

Capra Ibex Linn:
Le Bouquetin
Der Steinbock
The Wild Goat

Antilope Recticornis Erxl:
La Gazelle du Bezoard.
Die Gaselle Bezoarbock.
The Aegyptian Antelope

Antilope Strepsiceros Pall:
Le Condoma
Das Afrikanische Kututhier.
The Striped Antelope.

Antilope Orientalis Erxl:
Le Tzeiran.
Der Hirsch=Bock.
The Chinese Antelope.

Antilope Gnou Tulbagh.
Le Gnou
Der Capsche Gnou.

Ovis Laticaudata Erxl:
Le Mouton de Barbarie.
Das Arabische Schaf.
The broad-tailed Sheep.

Bos Bubalis Linn:
Le Buffle.
Der Büffel-Ochs.
The Buffalo.

XXXIV.B.

Bos Indicus Linn:
Le Zebu.
Der Zwergbüffel.
The Indian Bull.

Equus Zebra Linn:
Le Zebre.
Das Africanische Tyger Pferd.
The Zebra.

Hippopotamus Amphibius Linn:
L. Hippopotame.
Das Nilpferd. Behemot.
The Hippopotame.

Sus Babyrussa Linn.
Le Babirousa.
Der Hirsch-Eber.
The Indian Hog.

Hydrochaerus Tapir Erxl:
Le Tapir.
Das Antothier.
The Longnosed Tapir.